The Natural History Museum

Animal Close-Ups

Birds

and other flying animals

Barbara Taylor

PETER BEDRICK BOOKS

McGraw-Hill
Children's Publishing

A Division of The McGraw-Hill Companies

Published in the United States in 2003 by
Peter Bedrick Books, an imprint of
McGraw-Hill Children's Publishing,
A Division of The McGraw-Hill Companies
8787 Orion Place
Columbus, OH 43240

www.MHkids.com

ISBN 1-57768-961-5

Library of Congress Cataloging-in-Publication Data is on file with the publisher.

Text copyright © Barbara Taylor 2003
Photographs copyright © The Natural History Museum, London 2003
Photographs by Frank Greenaway

The moral rights of the author have been asserted

Database right Oxford University Press (maker)

1 2 3 4   5 6 7 8 9 10 OXF 06 05 04 03 02

Printed in Hong Kong

Contents

We can fly!

Only three kinds of animals can fly: birds, bats, and insects. Animals fly to escape from danger and to find food and safe places to rest.

Birds are the fastest and most powerful fliers. I am a hawk. My wings are curved on top and hollow underneath. This is a good shape to lift me up into the air.

I am a furry fruit bat. I am a mammal, like you. Bats are the only mammals that can fly.

I am a flying squirrel, but I cannot really fly. I glide using flaps of skin along the sides of my body.

I am a wasp. Like most insects I have two pairs of wings. Their network of veins makes them strong.

What are wings made of?

I am a long-eared bat. My wings are made of leathery skin. They are supported by my finger bones.

I am a gliding gecko. The flaps of skin along my sides are a bit like wings. But I cannot flap them up and down.

I am
a hawk.
My wings are
made of feathers that
are joined to my arm
bones. My flight feathers fit
closely together, like a fan.

I am a dragonfly.
My wings are made of
the same material as my
hard body covering. I can
hover, dart backwards and
forwards, and stop suddenly.

9

I am a powerful eagle.

Using my long, wide wings I float high above the African wilderness. When I come in for a landing, my tail and wings spread out, and my legs swing forward to cushion the impact of landing.

My wing feathers steer me through the air.

I have a third eyelid on each eye that helps protect it.

My deadly talons are strong enough to crush and carry off my prey.

My strong chest muscles give me the power to flap my enormous wings.

I am a speedy hummingbird.

My glossy feathers change color in the light.

My wings beat so fast they make a humming sound.

My long, slender bill reaches deep inside flowers for the sugary nectar.

I am the only bird that can hover forwards and backwards as I feed from flowers.

12

I am a quacking mallard duck.

I am a strong flier and a good swimmer. I use my wide, flat bill to sift food out of the water.

The down feathers on my breast are soft and fluffy.

I use my webbed feet like paddles to swim.

I am a male. I use my colorful feathers to attract a female. She has streaky brown feathers.

13

I am a graceful swan.

I am a big, heavy bird. Before I fly, I run along the surface of the water flapping my huge wings. They then lift me up into the air.

I look very graceful on the water.

14

I have to clean my feathers every day to keep them in good condition. I spread oil on them to keep them waterproof.

I am a male, with a black bump on my bill.

My babies are called cygnets. They cannot fly until they have grown their flight feathers.

15

I am a furry fruit bat.

I eat ripe fruit. I use my hands and arms to fly. My fingers are very long.

I use my big eyes and long nose to find food.

I move my arms up and down to flap my wings. My strong chest and shoulder muscles help me to fly.

16

I hang upside down
with my friends and
family. We fold our wings
away while we rest.

Can you see my short thumbs?

My wings are made of leathery skin
that is stretched between my long finger
bones. My wings are light but strong.

17

I am a buzzing hoverfly.

I am an expert flier. I can hover, like a hummingbird, and move in any direction, even backwards.

I have only one pair of short wings. Their veins make them strong.

My tiny antennae sense the air around me.

I suck nectar from flowers through a long tongue, called a proboscis. It folds away under my head when I am not feeding.

My huge eyes help me to see well when I am flying fast.

19

I am a busy beetle.

My delicate wings are folded neatly away under the hard wing cases on my back.

My bulging eyes help me to spot danger.

I use my jointed feet for climbing and clinging.

20

I am a peacock butterfly.

I flap my four big wings and glide slowly through the air.

As I came out of my cocoon, I pumped blood into my wings to make them stiff and flat.

The tiny scales on my wings overlap, like roof tiles.

Glossary

antennae Long, thin stalks on an insect's head, used for sensing things.

down Soft, fluffy feathers that grow close to a bird's body and keep it warm.

flight feathers Large feathers that make up an adult's wing and are shaped for flying.

insect A small animal without a backbone, that has six legs and three parts to its body.

mammal An animal that feeds its young with mother's milk. Human beings are mammals.

prey An animal that is killed or eaten by another animal.

talons The strong, thick claws of a bird of prey.